Performance Scheduling

By

John Revere, PMP

ISBN: 1-4107-0960-4 (e-book)
ISBN: 1-4107-0961-2 (Paperback)

Library of Congress Control Number: 2002096920

This book is printed on acid free paper.

Printed in the United States of America
Bloomington, IN

1stBooks - rev. 05/21/03

Table of Contents

1. PERFORMANCE 1

2. BASICS OF SCHEDULING 3
- STATEMENT OF WORK 3
- PROJECT BREAKDOWN STRUCTURE - PBS 3
- WORK BREAKDOWN STRUCTURE - WBS 4
- ORGANIZATIONAL BREAKDOWN STRUCTURE – OBS 7
- TIME POSITIONING 8
- LOGICAL RELATIONSHIPS 9
- CALENDARS 12
- WORK EVENT TYPES 13

3. RESOURCE APPLICATION 19
- ROSTER 19
- AVAILABILITY 19
- UTILIZATION 20
- PRODUCTIVE & NONPRODUCTIVE 20

4. LEVEL OF DETAIL 25
- INTRODUCTION 25
- PROJECT SCHEDULE 25
- OPERATIONAL SCHEDULES 27

5. CRITICAL PATH ANALYSIS 31
- FORWARD PASS - EARLY START & EARLY FINISH 31
- BACKWARD PASS - LATE START & LATE FINISH 32
- EVENT FLOAT 34
- FREE FLOAT 35
- PROJECT FLOAT 36
- IMPOSED DATES 36
- MILESTONES & MAJOR EVENTS 37
- CRITICAL PATH IDENTIFICATION 37

6. ESTIMATING 41
- ESTIMATING RANGES 41
- ANALOGY ESTIMATING 41

- PARAMETER ESTIMATING 42
- DEFINITIVE ESTIMATING (ENGINEERING) 42
- BID SOLICITATION 43
- PERT ESTIMATING 43

7. RESERVES 47
- INTRODUCTION 47
- CONTINGENCY RESERVE 47
- MANAGEMENT RESERVE 48

8. PERFORMANCE MEASUREMENT 51
- EARN VALUE ANALYSIS 51
- PROJECT EARN VALUE ANALYSIS 52
- OPERATIONAL SCHEDULE ANALYSIS 61
- ISSUES CONCERNING EARN VALUE 66
- FLOAT ANALYSIS 68
- CRITICAL PATH ANALYSIS 68
- RESOURCE ANALYSIS 69

9. SCHEDULE IMPROVEMENT 73
- INTRODUCTION 73
- CRASHING 73
- FAST TRACKING 74

GLOSSARY 79

CPM SCHEDULING LESSONS 91

CPM PROJECT SCHEDULING 103

PERFORMANCE

1

INTRODUCTION

The importance of scheduling a project and then monitoring of its performance is a key element in the success of any project. No other management task is more demanding of a project team's time. The development of a project's schedule and its routine updating takes time and effort, which must be made available to the project team. The skills required for scheduling a project successfully are learnable and the proper application of scheduling a project can be framed into a logical order; that is what this text is all about. While this text is focused on the construction industry, the scheduling concepts are universal in their application to all industries. I focused on the construction industry for the basis of this book since typical construction projects involve: dynamic changes in scope, limited resources, tight budgets, multiple subcontractors, a wide assortment of risk elements, tight specifications, unreasonable contracts, a requirement for mass communications, the maintenance of high quality objectives, and most of all, a predefined completion date.

To be successful in today's marketplace, you must have a plan of action and must transform that plan into a schedule and then execute that schedule.

Time is money, and money is based on time.

1

John Revere

BASICS OF SCHEDULING

2

STATEMENT OF WORK

In the development and the monitoring of a project schedule, we begin with the project's Statement of Work (SOW). This brief statement describes the essence of the project. In this statement the project is described, at minimum, by the project's triple constraints, namely its anticipated duration, its projected cost and the anticipated quality of the project. With these elements of the project identified, the project team can begin to design the project schedule.

The project team must analyze the relationships between the project's triple constraints and other elements of the project to understand what will delight the customer at the project's completion.

PROJECT BREAKDOWN STRUCTURE - PBS

The Project Breakdown Structure, sometimes referred to as a Product Breakdown Structure, describes the deliverables of the project. It outlines all the major deliverables that will fulfill the objectives of the project as stated in the project's Statement of Work. In describing the deliverables of the project, a specific written description is made using nouns. This description describes what is to be delivered, not how it is to be accomplished. The triple constrains of the project, or of a deliverable, are not addressed at this level. Once the Project Breakdown Structure (PBS) has been identified, a unique numeric code is assigned to each deliverable. Descending sub-deliverables carry the master deliverable identifier so that a logical grouping of project deliverables is maintained. The project's management team determines the depth or level of detail of these deliverables. The level of detail of a deliverable is normally based on the project team's reasonable ability to monitor the execution of the deliverable.

On determining the level of sub-deliverables for the project, the project team must also think through addressing the project's Code of Accounts. The Code of Accounts establishes areas of the project that the financial benefactor of the project requires to detail or summarize all costs associated with a particular deliverable or an area of the project. Very often this accounting practice deals with the benefactor's plans to later segment tax depreciation on a project's deliverables. An example of an identifiable deliverable, which may require a unique account code, may be for a specific piece of HVAC equipment that has a short life expectancy. The project's benefactor may wish to specifically identify this piece of HVAC equipment for rapid depreciation/replacement purposes.

WORK BREAKDOWN STRUCTURE - WBS

The Project Breakdown Structure defines all the project deliverables; the Work Breakdown Structure describes how these deliverables are to be accomplished. While nouns were primarily used to describe the PBS, the WBS uses verbs to describe actions to be taken to satisfy the PBS. The WBS outlines how these deliverables of the PBS are to be executed. These executable portions of the WBS are referred to as work packages. Individual work packages are then assigned to a discipline or contractor who will be responsible and accountable for the work. Some of these work packages can then be grouped together for competitive bid purposes for subcontractor support on the project.

Work packages continue with the logical decomposition of numeric identification as predefined by the PBS. Once again the level of detail of these action items, work packages, are to be determined by the project's management team.

As work packages are assigned to individual organizations of the project, the work packages should then be further broken down into work activities that describe the actions required to complete a work package. This level of planned execution, work activities to satisfy the assigned work package, is the responsibility of the assigned organization - not the project management team. This work package breakdown forms an Operational Schedule for the assigned organization.

The Operational Schedule of an organization may stop at this work package activity level as far as reporting to the project management team as to the assigned project's work package performance status. Many organizations will elect to subdivide these operational work package work activities into an operational task schedule so that the individuals working on the activity will have a step-by-step procedural guideline to insure quality control.

John Revere

<u>*Outline Example of PBS, WBS, Activity, Task Levels*</u>

2.0	MECHANICAL SYSTEMS	PBS
2.1	Sprinklers Systems	PBS
2.2	Plumbing Systems	PBS
2.3	HVAC Systems	PBS
2.3.1	Unit 2A - North Wing	PBS
2.3.2	Unit 2B - North Wing	PBS
2.3.2.1	Deliver Unit 2B to Site	WBS
2.3.2.2	Install Pneumatic Lines	WBS
2.3.2.3	Install & level HVAC Unit	WBS
2.3.2.3.1	Bring unit 2B to Location	Activity
2.3.2.3.2	Set and level Unit	Activity
2.3.2.3.2.1	Remove Unit From Shipping Box	Task
2.3.2.3.2.2	Install Unit on Pad Foundation	Task
2.3.2.3.2.3	Install 5 CER Isolators	Task

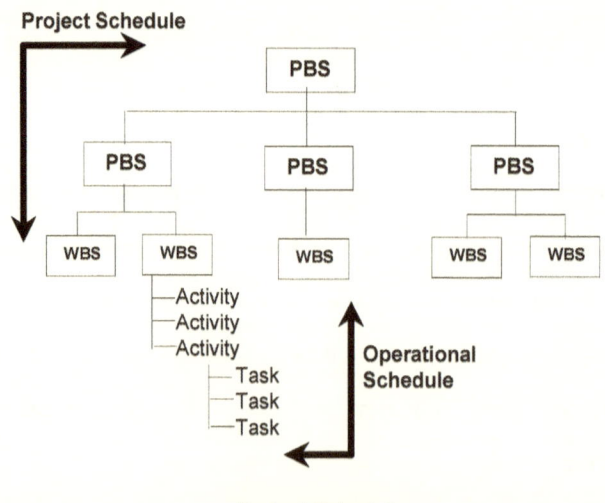

Project Schedule
(PBS, WBS, Activity, Task)

ORGANIZATIONAL BREAKDOWN STRUCTURE – OBS

Work packages of the project schedule are assigned according to responsibility and accountability for their completion. An organizational chart profile of all project members is compiled for this effort by the project management team. This chart, called an Organizational Breakdown Structure (OBS), stipulates existing and future organizations, which the project requires.

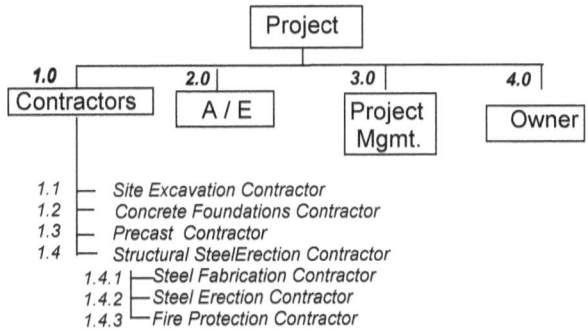

Organizational Breakdown Structure - OBS

The OBS is developed in a manner similar to the PBS.

Initially, the major players of the project, including the owner, are identified along with major anticipated contractors. As the project matures and bids are awarded, the OBS is updated.

As work packages are identified in the project schedule, they are also matched to the OBS. This matching of WBS and OBS is referred to as a Responsibility Assignment Matrix, or RAM.

TIME POSITIONING

Today most if not all project scheduling software programs are based on a precedence method of scheduling - PDM. With PDM there are four major time positioning types which allow the project scheduler to position work packages and work activities into the project schedule.

ASAP - As Soon As Possible

This is the most commonly referred to positioning type, and is used as a default by most scheduling programs. It requires the work package and its work activities to be completed "as soon as possible."

ALAP - As Late As Possible

This positioning type is not commonly used in scheduling since this would require the work package and its work package activities to be accomplished "as late as possible" in the project schedule. This positioning type usually increases risk to the project since it automatically eliminates the activity float assigned to the work, and therefore places the work to be completed on the critical path of the project.

Must Dates

A "must date" can be imposed on any work packages and its work activities. It is a hard date, established by the project team, that can be imposed on a start date, end date, or both. Some software scheduling tools allow for a "must start by" date or "must finish by" date. The use of must dates should be directly identified in the scope document, which was established during the development phase of the project. The use of "must date" should be kept to a minimum since they often violate the logical flow of work assignments.

Work Between

This time positioning type allows the scheduler to establish an early start and a late finish date of a work package or work activity while allowing for the work to float between these dates. The work is elastically stretched between these two points in time. This allows the scheduler flexibility in dealing with project events that extend from a point in time to a specific end date or period. Management, administrative and other overhead project work packages normally use this form of time positioning.

LOGICAL RELATIONSHIPS

The most commonly used method of scheduling today is Performance Diagramming Methodology - PDM. Precedence Diagram Methodology is a methodology that clarifies the relations between work events using logical relationships between the work events. Precedence Diagramming Methodology identifies four basic logical relationships that are used to describe one work event (work package or work activity) with another work event. These four relationships are: finish to start, start to finish, start to start, and finish to finish relationships.

Finish to Start

In a finish to start relationship (FS) work event, A must be completed in order for activity B to start. This is the most commonly used relationship in scheduling project work events.

Start to Finish

In a start to finish relationship (SF), after work event A has started then work event B can then finish. *(opposite logical work flow from FS - not often used in scheduling)*

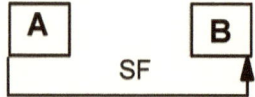

Start to Start

In a start to start relationship (SS) after work event A begins then work event B can begin.

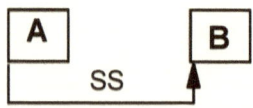

Finish to Finish

In a finish to finish relationship (FF) after work event A finishes then work event B can be completed.

Relationship Realism

The logical relationships between work events as defined in the PDM relationship are upgraded by the use of lags and leads in association with these logical PDM relationships. A lag is a wait period or a time delay between work events. A lead is an overlap period of time between work events.

Lag Example:

Paint is required to dry 10 hours before artwork can be hung.

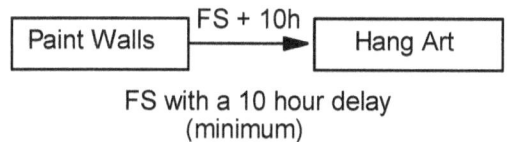

FS with a 10 hour delay
(minimum)

Lead Example:

The painting of all building interior walls will take 5 days. Artwork for the building will take 10 days to hang. It has been decided by the project team that there can be a 2 day overlap (lead) between painting and hanging of artwork.

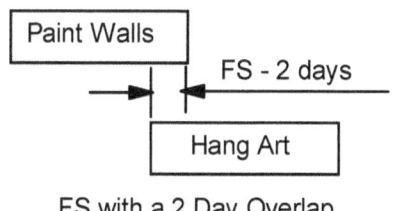

FS with a 2 Day Overlap

CALENDARS

Project Calendar

The highest level of calendars is the project calendar. This default calendar is used by the project scheduler to determine the project completion date in which major holidays and non-work periods have been identified. The establishment of this calendar is based on agreement with the project team and the owner/client.

Project Calendars may include defining: non-work periods like all Saturdays and Sundays, owner imposed non-work periods, etc.

Group Calendars

Group calendars are assigned to groups that share the same work patterns or calendars. These group calendars may supersede project calendars. Grouping of resources normally takes place when a particular trade or subcontractor is assigned work and these groups operate with different work hours and /or different time-off periods.

Work patterns of group calendars may include: non-work periods for union holidays, department imposed general meetings, boss imposed times, legal break periods, highway movement restrictions, etc.

Resource Calendars

Resource (human and non-human) calendars are assigned to a specific entity working on the project. These individual calendars identify a unique work period that may supersede group or project calendars.

Resource calendars may identified: specific work hours, vacation periods, training, jury duty, the rental periods of equipment, etc.

WORK EVENT TYPES

The work to be accomplished on any event can vary depending on the effort required. It is imperative that those scheduling a project have a solid understanding of each of these work type selections and their applicability. It should be noted that, depending on the computer scheduling software being used, there can be differences in definition and availability.

Effort Driven Event

An effort driven work event is one in which additional resources will directly affect the duration of the event. By adding more resources, the duration of the event is decreased. In the example below, the total resource hours equals 20 hours and the duration of the work event of 20 hours is based on one resource being applied; however, the duration can be reduced from 20 hours to four hours with the assignment of five resources to the work event. In an effort driven work event, the event resources hours are fixed, but the work event duration can fluctuate. The efficiency of resources and the environment of the work event directly impacts the calculation of a work event's duration.

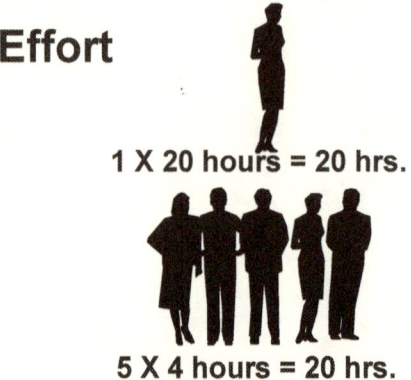

Effort

1 X 20 hours = 20 hrs.

5 X 4 hours = 20 hrs.

Workday Driven Event

A workday event is often referred to as a fixed duration event since the event's duration is set. The number of resources applied to the work event has no impact on its duration. Adding additional resources to a workday driven event increases the total resources hours associated with the work event, but does not alter the work event's duration.

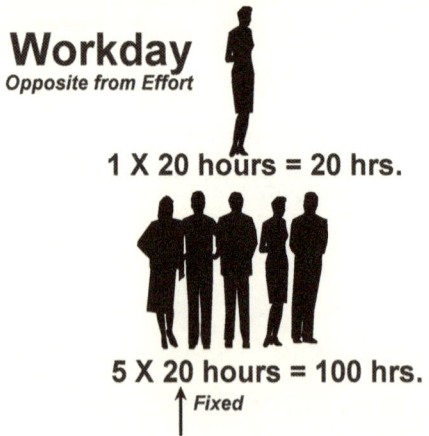

Workday
Opposite from Effort

1 X 20 hours = 20 hrs.

5 X 20 hours = 100 hrs.
↑ *Fixed*

Elapsed Duration Event

An elapsed duration event is similar to the workday driven event since the duration of the event is fixed. The difference between these two types of event is that the duration of an elapsed event is established, even though the resources assigned to the event may not be required for the entire duration of the event. In the example below, resources (painters) are required to work Thursday and Friday only, allowing the painted room to dry three days, thus creating an event with a duration of five days. Depending on the computer scheduling software, an elapsed event may also ignore project and group non-work calendar periods.

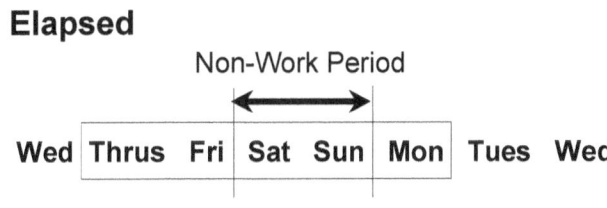

Elapsed

It takes five days the painted room to dry.

Resource Driven Event

This event type is the most commonly used type in project scheduling since individual resources directly assigned to the activity determine the duration of the work event. Using the example below, three resources have been assigned to a work event. The work event duration is based on the middle resource, which plans on spending three days on the work event.

Resource Driven

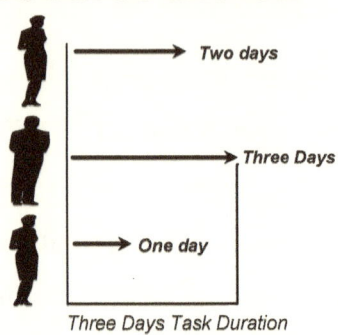

Two days

Three Days

One day

Three Days Task Duration

Span Work Event

The span work event does just that, the event spans over a period of time. The span or duration of the event depends on its relationship with its predecessor and successor work events. The work event expands or contracts in duration depending on these relationships. In the example below, the work event B is the span work event.

Normally, span work event are used in preliminary project schedule development when there is a lack of detailed information about the work, but a specific performance period has been established. Span work events are generally used in these preliminary schedules until more detailed scheduling information becomes available to the project management team. The use of these event types may aid in the establishment of preliminary schedule objective dates.

Span Event

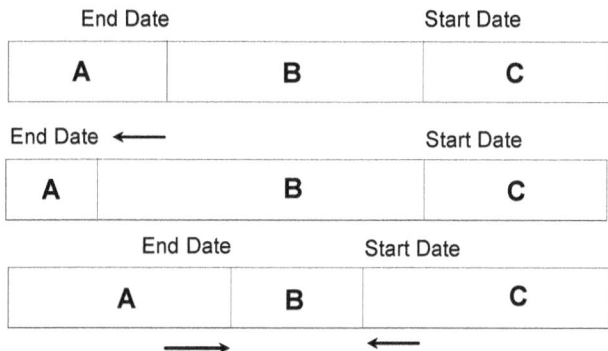

John Revere

RESOURCE APPLICATION

3

ROSTER

The OBS (Organizational Breakdown Structure) identifies which organizations are required for the project. Each organization may be subdivided into functional areas of its organization. The *Roster* is the total number of resources that an organization has identified as workers in a functional area of the organization. The roster may be broken down into skill levels or areas of specialty, this is determined by the management of the organization. The roster may be broken down by product lines, divisional location of workers, or by craft trade. The more detailed the roster is, the greater the control the organization will have in its management of the project and of organizational resources.

AVAILABILITY

The *Availability* of workers is the Roster minus deductions for typical employee benefits such as vacation, sickness, time-off, off-site training, jury duty, maternity leave, and authorized company leave. Availability rates vary with each organization, in a union environment this rate can average between 12% to 18%, in a nonunion technical trade environment this can average from 8% to 15%. It is therefore necessary for the project team to identify these organizational rates with the human resources department of the project organization.

UTILIZATION

The *Utilization* is the Availability minus deductions that remove these available resources from the project. Deductions normally include lunch period, scheduled breaks, setup time, authorized union meetings, boss imposed time, coordination of work assignments, coordination of the work, meetings, and on-the-job training. Once again, these rates vary depending on the project work environment and more importantly the work environment of the organization. Utilization rates can vary between 5% in a nonunion technical environment to 12% in a union craft environment.

PRODUCTIVE & NONPRODUCTIVE

After the utilization rate has been factored into the resource loading, the resulting factor is divided into two classifications of investigation planning, namely Productive and Nonproductive. *Productive resources* are those resources of the utilization that are associated with work that is to be performed on the project. *Nonproductive resources* are those resources that are scheduled to work on the project but are not doing so. Nonproductive activities include miscommunication, inefficient work environments, poor equipment usage, lack of tools, community /social relations. It should be noted that nonproductive resources are often mistaken as being detrimental to an organization and to the project. However, many organizations desire the worker to believe that its organization is more than just a place of work - its a workplace that believes some nonproductive time is essential in creating a positive work environment. Nonproductive and productive resources in a unionized craft environment typically averages 85% productivity and 15% non-productivity of the utilization total, while in a nonunion technical environment, this range can be 70%-80% productive and 30%-20% nonproductive.

Productivity is a rate of how efficiently workers work. Productivity is measured on productive workers (excluding the nonproductive) only. It's based on the idea that a certain number of productive workers should produce a certain quantity of work units. To measure productivity, an established standard of performance needs to be created. For some industries, like the construction, these industry these standards have been established, for other areas, like high technical design, standards are still being formulated. With established productive resource hours for project work events, actual performance (productivity) can be measured. Project management and those supplying resources for the project must realistically address each of these resource subjects. Those in administration monitoring productive and nonproductive rates must be aware of seasonal fluctuations and adjust the project application rates of resources accordingly. Those organizations supplying resources must be cautious when attempting to reduce nonproductive rates beyond what labor believes are justifiable.

An attempt to reduce nonproductive resource hours often results in an increase in resource hours or poor productivity performance. A friendly, positive work atmosphere, where the individual manages nonproductive time, often results in increased productivity within the organization and thus benefits the project.

Resource Identification
* Range Varies 2%-5%

Roster	Availability	Utilization	Productive & Non-Productive
	Physically Not At Work	Legal or Contractual Agreement	
100	82	72	61 — Productive 85%
	18%	12%	11 — Non Productive 15%
	Deduct Vacation, Authorized Leave, Sickness, Training off Site Jury Duty, etc.	Deduct - Breaks, Lunches, Set-up Union Meetings, Coordination, Parts, OJT, etc.	*Non-Productive Areas* Social Environment, Waste, Mis-communication, Down time inefficient work environment, poor equipment usage. etc.

WORK

John Revere

LEVEL OF
DETAIL

4

INTRODUCTION

The level of detail one should schedule for a project depends on the amount of control that the scheduler contemplates on having during the project's life cycle. Many organizations have established "core" schedules that have predefined levels of project schedule detail in which unique project specific events are then added by the project team. The development of "core" schedules greatly enhances the speed, accuracy and quality level of scheduling projects. Care must therefore be taken when developing these "core" schedules since any error in logic, resource application, or scope of work can have a cascading effect not only on the project but on the entire organization.

PROJECT SCHEDULE

The level of detail for a project schedule can vary depending on a number of factors, among those are: the project environment, time for the development of the project schedule, the duration of the project, the complexity of the project, the financial code of accounts, the size of the project, and which group of the OBS are going to be responsible for the work package. In a project schedule, we do not schedule work activities and work tasks since these are elements of an operational schedule. In a project, we schedule to the event level of the WBS, the work package that describes how the deliverables (needs) of the project are to be completed (how). Depending on the factors imposed on the project scheduler, the level of detail will be once again based on the planned control update procedures of the project.

Typically on projects, schedulers on their first draft of a schedule develop the schedule at the lowest level of the project, and after identifying the project's critical path, revisit the project's critical path work packages to develop a more detailed understanding of those work packages.

Example:

Building Foundations (need) has been identified as part of the PBS of a project. The WBS breaks deliverables down into work packages (how this work is to be accomplished), which will then be scheduled as part of the project schedule.

General Pass: *Install foundation concrete wall footings.*

Install foundation masonry walls.

Detail Development:

Install foundation concrete North wall footings,

Install foundation concrete South wall footings,

Install foundation concrete East wall footings,

Install foundation concrete West wall footings;

Install foundation North masonry wall,

Install foundation South masonry wall,

Install foundation East masonry wall,

Install foundation West masonry wall.

OPERATIONAL SCHEDULES

Operational schedules take the WBS work packages, defined by the project schedule, and breaks these packages into work activities that describe the "actions" required that must take place to complete the work package. An activity may be further broken down into work tasks that describe the "procedure" required to complete the work activity. The project scheduler rarely works at this level; this level is focused on those in functional management who are directly responsible for the work that is to be completed. Functional department managers, on-site supervisors, contracted help, or those actually doing the work develop the activities and tasks associated with the operational schedule. The level of detail of these activities or tasks is dependent on the same environmental factors that project schedules have, not enough time and a lack of scheduling knowledge being amplified at this level.

There are numerous rules concerning the size of operational activities and tasks, which the management of the operational schedule needs to address. Some of these rules address the maximum number of resource hours that are not to be exceeded, i.e. 40 hours maximum per work activity. Some organizations have established maximum limits for this development; i.e. that no work task is to be assigned to more than one resource, or that no activity is to exceed 10 days in duration. .

The rule concerning the size of work activities or tasks reverts back to the level of control expected during the operational life cycle. The level of detail in the operational schedule needs to be agreed to between the responsible operational manager and the employee completing the work for that functional area.

EXAMPLE

WBS Work Package (Detail)   *Install foundation concrete North wall footings.*

Work Activities: *Establish footing elevation.*
Set side-footing sideboards.
Verify mech. and elect. Placements
Prep footing base with rock.
Pour, tamp and level concrete.
Install footing key-way.
Install wall-reinforcing rods.

Work Activity: *Establish footing elevation.*
Work Tasks: *1. Verify site elevation point.*
2. Calculate footing elevation and record this finding.
3. Measure new footing elevation measurement (recheck).
4. Establish elevation footing stake. (recheck)
5. Transfer this to each footing wall elevation stakes.

John Revere

CRITICAL PATH ANALYSIS

5

FORWARD PASS - EARLY START & EARLY FINISH

Once the logical network has been developed by the project team, a forward pass is then made through the network. This forward pass is a mathematical calculation of the work event duration, plus any lags or leads associated with the logical sequencing of the work events. To do this calculation, one begins at the start date of the project (if no date has been established, use zero) and moves toward the end work event of the project. With this earliest project start date, add the event's duration which determines the event's earliest finish date for the event, and continue moving toward the project's last event. The example below starts at the beginning of the project, using the identified early start date of the project, which is "zero".

Early Start date + event duration = Early Finish date of the event.

Forward Pass Example

   Starting with Work Event A, the earliest Event A can start is point zero and the earliest it can finish is day 2. Once Event A is complete, then Event B & C can begin. The earliest Event B can finish is day 6, and day 5 for Event C. The early start of Event E is 6, based on the completion of Event B, and its early finish day 10. The earliest Event D can begin is day 6, based on Event B's completion, its finish day being day 8. The project's final event, Event F, starts on day 10 based on the completion of Event E. The project's earliest possible finish date is day 12.

Note that non-work periods like weekends are not to be introduced at this time or any other non-work period restrictions to the project.

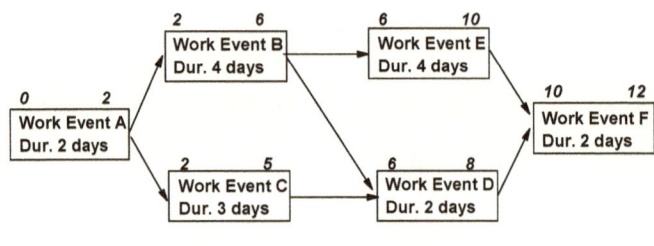

Forward Pass
Early Start & Early Finish

BACKWARD PASS - LATE START & LATE FINISH

Once the project team has calculated the forward pass, a backward pass is then made through the network. This backward pass is a mathematical calculation of the work event duration, minus any lags or leads associated with the project. To do this calculation, begin with the last work event's latest finish date and move towards the first event of the project by subtracting the event's duration, which will identify the Late Finish and Late Start for each work event of the project. Under ideal conditions the project scheduler would use the early finish date of the last work event, which was calculated using the forward pass as the completion date of the project, however under normal conditions executive management or the client may have predetermined the project end date.

In the CPM example, next page, start this analysis with the calculated project finish date (Early Finish) for the project, or the calendar finish date if identified, and then working backward, subtract each event's duration along with relationship lags or leads.

Late Finish date - event duration = Late Start of the event.

Backward Pass Example

Starting with Work Event F, the latest Event F can finish is day 12 and therefore the latest it can start would be day 10. Moving toward the first work event of the project, Event E Late Finish date is equal to the Late Start of Event F, and subtracting the Event's duration from the Late Finish of 10 gives us a Late Start of 6 for Event E. Event D Late Finish is equal to 10 and subtracting the event's duration from the Late Start of Event D becomes day 8. Event C's Late Finish date is 8 based on Event D's Late Start, and the Late Start of Event C is 5 subtracting its duration from the Late Finish date. The work Event B's Late Finish is based on the earliest date possible of the Late Start dates of Event D or E. In this example, select the earliest start date, day 6, from Event E to be the Late Finish date for Event E. The Latest Start for Event E is day 2 calculated by subtracting the event's duration from its Late Finish date. Finally the Late Finish date for Event A is based on the earliest Late Start date between Event B & C, in which we select day 2 from Event B. The Late Start for Event A is day zero based on the Late Finish minus Event A's duration of two days which equals day zero.

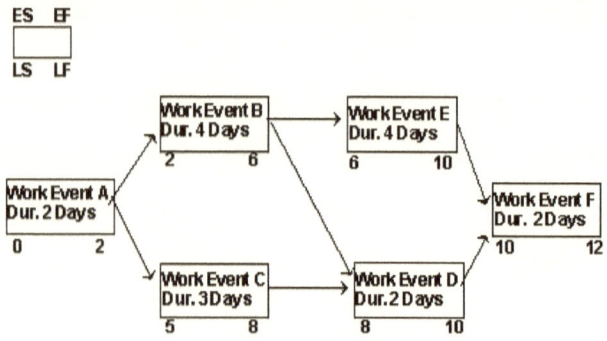

Backward Pass
Late Start & Late Finish

EVENT FLOAT

Float, sometimes referred to as "slack," is the amount of time an event can move in time without having a negative influence on the project's completion date. The difference between the Late Start of an event and the event's Early Start determines the amount of float an event has. Since the duration of an event is fixed, one can also calculate float by observing the difference between the event's Late Finish date and its Early Finish date.

Under ideal scheduling conditions, project teams attempt to complete work events as soon as possible, using the Early Start and Early Finish dates. There are at times (under normal conditions) that resources limitations, material shortages, and other conditions hamper the Early Start dates, which are not then conducive to the project demands. The project team then reviews the float of the work event in order to find an acceptable date of execution of the event without going beyond the Late Start date of the work event.

Late Finish - **Early Finish** = Float

or

Late Start - **Early Start** = Float

Event Float Example - (See previous Examples)
The Work Event C, has an event float of 3 days.

LF (8) - EF (5) = 3 days of event float

or

LS(5) - ES (2) = 3 days of event float

FREE FLOAT

Free float is not calculated on a work event's start or finish dates. Free float is based on a logical relationship between two work events. The precise calculations of relationship free float is calculated by observing the difference of an event's Late Finish date and that of the event's successor event Early Start date. Free float is relationship based.

Late Finish (event) **-** **Early Start** (succeeding event) = Free Float

Free Float Example
The Work Event B to Work Event D, has a relationship free float of 0 days.

LF of event B (6) - ES of event D (6) = 0 days of event float.

PROJECT FLOAT

Project float, sometimes referred to as Total Float, is the difference between the calculated Early Finish date of a project based on the project events, and its imposed or calculated Late Finish date. Imposed dates may be determined by marketing requirements, the client, or by contract agreement. Normally, the imposed latest finish date of the project becomes the Late Finish date of the last event of the project. The difference between the imposed Late Finish date and the project's Early Finish date is the project's float period.

Late **F**inish (imposed date) **- E**arly **F**inish (calculated)= Project Float

Project Float Example
The Work Event F is the final work event in our project schedule. The Work Event F late finish to Work Event F's early finish date has a project float of 0 days.

LF of event F (12) - EF of event F (12) = 0 days of project float

IMPOSED DATES

There may be imposed dates within a project schedule. These dates are often imposed by the project team to meet a contract or client selected date. By imbedding these dates into the project schedule, the critical path usually is usually affected. The use of imposed dates should be kept to a minimum so that the true logical sequence of the work of the project can be determined and managed without undue manipulation.

MILESTONES & MAJOR EVENTS

A *milestone* is an event that has zero duration and can be directly related to the Scope of the Work (SOW) of the project. Milestones are often itemized within the project's contract requirements. Since these dates are imposed dates on the project schedule, their application should be limited.

A *major event* is an event that has a zero duration, similar to a milestone, however it cannot be traced back to the SOW. Major events are often referred to as boss imposed dates since these events may be important to a management segment, usually upper management, that requests these dates be placed into the project schedule to highlight an event that is important to them, but not the project.

CRITICAL PATH IDENTIFICATION

Those events that have the least amount of event float are critical to the project's completion. These events are critical because their Early Start date and Late Start date are the same date, thus having zero event float. Those work events with the least amount of float are linked together in the project network which forms a path - a critical path. The critical path forms the longest path through the project network. This path also determines the shortest time in which the project can be completed. A positive deviation in any of the critical work events will cause the project to be done ahead of the originally planned completion date. If the original critical path events increase in the amount of positive float, the critical path of the project is more than likely to move to a new location/path within the project network. A negative deviation in any of the critical work events will cause the project to go later than the originally planned completion date.

It is of the utmost importance that the project team focuses on the critical path work events and those events near the critical path to maintain schedule objectives.

John Revere

ESTIMATING

6

ESTIMATING RANGES

There are three different project estimating ranges: order of magnitude, budgetary, and definitive. These ranges express a degree of accuracy of the estimate being given.

Type	*Range %*
Order of Magnitude	+ 75% to - 25%
Budgetary	+ 25% to - 10%
Definitive	+ 10% to - 5%

When compiling estimates, one uses different methodologies or a combination of estimating methodologies. The most commonly used methodologies include: analogy, parameter (mistakenly called paramctic), definitive or engineering estimates, bid solicitation and PERT (program evaluation review technique).

ANALOGY ESTIMATING

An analogy estimate is based on expert opinion. The number one beneficial use of this type of estimating is that it's quick. The disadvantage of this methodology is that it is highly risky. Analogy estimates should only be used when there is little information available or when time is limited.

Example: To build a house, plan on spending $150,000, based on an expert opinion of a local builder.

PARAMETER ESTIMATING

Parameter deals with mathematical relationships. The use of this estimating methodology is based on having historical information that a comparison can be made. This estimating methodology is more accurate then the analogy method since it utilizes similar project historic data, however, caution should be used making sure that this data is not stale data. Parameter estimates take more time then the analogy estimate since similar project data must be compiled prior to the event being estimated.

Example: A recently completed 10 mile highway costs 2 million dollars, a newly proposed 15 mile highway is estimated to be 3 million dollars.

DEFINITIVE ESTIMATING (ENGINEERING)

Definitive estimates are developed at the lowest level of possible investigation of the work event. Those who will be responsible for the work or have expert knowledge of the work required to complete the work event usually develop this estimate.

Example: According to our lead carpenter, a typical A-1 wall is $2.89 per linear foot that is based on: wall studs at $1.29, gyp.wallboard at $.40, painting at $0.10, electrical $1.10.

BID SOLICITATION

Getting an actual quote is a great way to determine what an event will cost. To obtain a quote, the project team is to clearly identify the work in which a bid is requested, this takes time and effort, even if sole sourced. Competitive bid taking and the analysis of these bids takes even more time and effort than a sole source quote but offers the project team a more accurate bid estimate of the work.

PERT ESTIMATING
(Program Evaluation Review Technique)

This form of estimating takes under consideration a statistical evaluation of estimated ranges of project performance. The use of PERT estimating provides the estimator a more accurate estimate of the duration or expected cost of the work event. While this estimating methodology provides greater accuracy, it is time consuming since three and not one estimate is required for PERT analysis.

Optimistic Estimate (OE) = *based on past performance records of similar maintenance projects with normal labor requirements, the minimum duration or cost.*

Pessimistic Estimate (PE) = *based on past performance records of similar maintenance projects with normal labor requirements, the maximum duration or cost.*

<u>Most Likely Estimate (MLE)</u> = *based on past performance records, current performance standards, labor situations, equipment, current technical skills, etc., the most likely estimate of duration or cost is made.*

$$\text{PERT Estimate} = \frac{OE + 4 \ (MLE) + PE}{6}$$

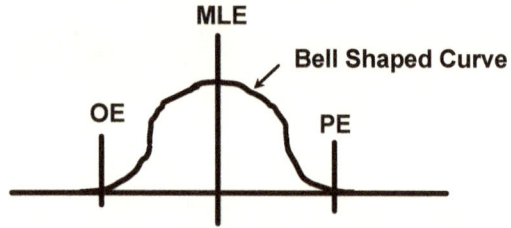

PERT Example

The cost of a work event has been estimated to be $5,00.00. A PERT estimate has been requested in which the following data was obtain: Optimistic estimate $4,000, most likely $5,000 and pessimistic $8,000. Based on this information a PERT estimate of the work event has been calculated to be $5,333.33

$$\text{PERT} = \frac{\$4,000 \ (Op) + 4 * \$5,000 \ (MLE) + \$8,000 \ (PE)}{6}$$

$$\text{PERT} = \frac{\$32,000}{6}$$

$$\text{PERT} = \$5,333.33$$

This estimate is $333.33 more than the original estimate given to the project (most likely estimate).

John Revere

RESERVES

7

INTRODUCTION

The use of reserves on a project is vital to a project's success. A reserve is to protect the project from known and unknown circumstances that could take place during the project's life cycle that would undermine the goals and objectives of the project. A reserve is not project fat, fluff, or is it a cushion, it is a management element of a project to correctly identify and mitigate risks in time, and cost and/or resources to increase the positive outcome of the project.

CONTINGENCY RESERVE

In any estimated work there is a known unknown factor that could impact the results of projected performance. A known factor is a factor the estimator knows that if it does occur, it could impact the work, however the probability of the factor remains unknown at the time the estimate is prepared. Contingency reserve identification is best defined at the lowest level of the estimate, normally this occurs at the task level, which is then accumulated at the activity level and then finally summarized at the work package level of the WBS.

A contingency reserve for a work package, which is planned for but not actually used is to be identified and set aside. The set aside contingency work package reserve may be identified as profit or as funds to be given back to the organization, however it is not to be used as a funding source for the project team for general or supplemental work, or to support other project work packages.

The percentage of contingency to be added to an estimate should be based on the estimator's experience in completing the work coupled with its anticipated work environment. Established organizational standards for work contingencies may be substituted for unacquainted work performance however, in using these standards the estimator must understand the environment in which these estimated standards were produced so that adjustments to the contingency reserve estimate can be made.

Example: A carpenter knows he can erect an eight foot long, 8 foot high, 2x4 wall in 15 minutes. His estimate for a 24-foot long, 8-foot high, 2x4 wall is 45 minutes, plus 15 minutes of contingency for a total of 60 minutes. The carpenter has added a contingency of 15 minutes to this estimate since the work environment will be outside winter construction were productivity will be lower then that of summer (normal) conditions.

MANAGEMENT RESERVE

Management reserve deals with the unknown elements that all projects have that could negatively impact the project. The project management team attempts to identify all potential negative project risk impacts but due to time constraints and limitations of knowledge about the work event, this simply is not possible, therefore the requirement to develop a management reserve for the project. Management reserve is often alluded to as funding to cover management mistakes in not identifying potential negative risk elements of a project. The reference to management reserve covering management mistakes is misleading since it implies it is to conceal the mismanagement of the project or poor performance, this is not the intent of management reserve. Management reserve funding is a conscious decision by the project team not to investigate all possible negative impacts to a project, usually due to a lack of time to investigate all possible project risks.

How much management reserve should be affixed to a project depends on the complexity of the project, the time allotted for project completion, anticipated quality of work, overall funding, other indirectly related project risks, but most of all the time required to develop the overall project time and cost estimates.

Example:

None exists (management reserve is an estimated fund for possible negative impacts of a project to cover unknown unknowns, therefore by its definition, nothing can be identified because then it becomes a known)

John Revere

PERFORMANCE MEASUREMENT

8

EARN VALUE ANALYSIS

The collection of unbiased data is critical if management is to analyze project status effectively. It is from this analysis that management judgments are formulated to bring expected future performance of the project into line with the project baseline plan. Management analysis is a process of comparing actual performance, analyzing variances, evaluating possible alternatives, and then taking appropriate corrective action as required. Applying the concepts of Earn Value Analysis is a popular performance measurement tool used by project managers since it measures project performance at any point in time within the project's life cycle. To apply the concepts of Earn Value, the project management team begins with the project schedule, those supplying resources for the project would begin with their operational schedule, which results would then transfer up into the project schedule. In either Earn Value Analysis, project or operational, the earn value application concept basics are the same.

Warning! *There are a number of textbooks and some professional organizations that do not understand the application of Earn Value. Their focus is on "generalization summation" of earn value measurement which summarizes detail performance information about the project and makes generalized statements about the project's future performance. To correctly apply earn value measurement principals on a project, you must analyze data at the lowest level of the schedule, the level of control of which data is gathered. Earn Value Analysis and estimates of future performance on time and cost are done at this lowest level, which then may be totaled to a higher level within the project.*

PROJECT EARN VALUE ANALYSIS

Project Earn Value analysis is based on taking a snapshot of the project at a point in time, normally this is done at the periodic project update-meeting period. Earned value methodology is a project performance measurement analysis. It compares the amount of work that was planned with what was actually accomplished to determine if there are cost and schedule variances. It is the project management team's responsibility to calculate the measurements of project performance and take corrective action in the best interest of the project and that of the client/owner.

It is important that the project team understands the basic concepts of Earn Value for correct application. The first step in Earn Value analysis is to review the existing project work packages. The scheduled work packages containing all work and cost associated with a project have a predictive range for a start and completion based on their logical relationships with each other. From this information expenditure forecast pattern over the life of the project can be calculated, this forms our project baseline. This baseline forecast of budgeted cost of work scheduled ((BCWS / PV)) forms the foundation of our project earn value analysis, sometimes referred to as a project's Planned Value (PV).

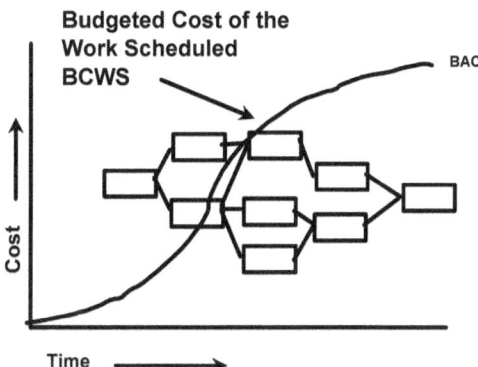

If the project team was to take a snapshot anytime during the project life cycle of the project they would able to identify the total cost of the work that should have been completed at that point in time ((BCWS / PV)) or planned value (PV) and compare it with the budgeted cost of the work that has been performance ((BCWP / EV)) or sometimes referred to as its Earn Value (EV). The earn value of those completed work packages would receive full value of the work package. Those work packages that are 30% completed, based on the project update, would have an earn value equal to 30% of its total value. The differences between what was to be completed and what has been earned would be a work package schedule variance (SV). Earn Value Analysis is to be performed at the lowest level of the project schedule - the work package. Project Earn Value analysis should never be performed at a summary project level, this leads to the summation generalization. Summation generalization is taking the current summary project analysis information, which is based on current performance, and based on that information make future performance estimates on future unrelated work packages and work packages that have not started.

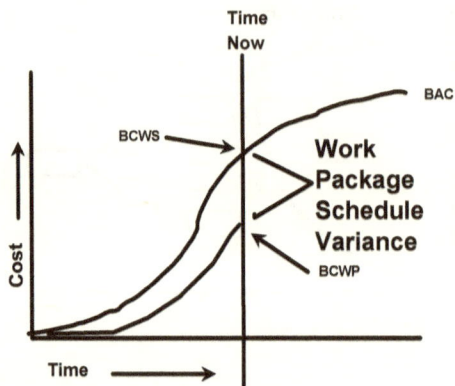

(BCWP / EV) work package - (BCWS / PV) work package = SV work package

A review of the difference between the budgeted cost of the work performed ((BCWP / EV) work package) and that of the actual cost of the work performed ((ACWP / AV) work package) would give the project management team a cost variance of the current work performed at this point in time. (CV work package).

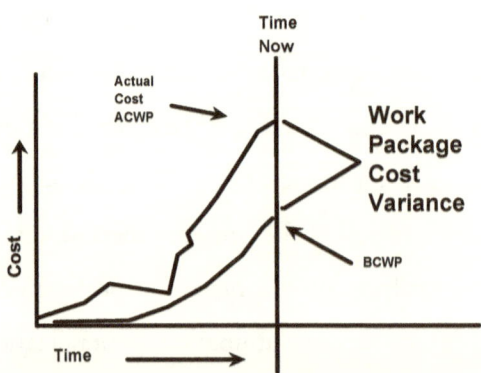

(BCWP / EV) work package - (ACWP / AV) work package = CV work package

From this information on cost and schedule variances, the project team can establish ratios on which estimates of the future cost and schedule can be based. One of these ratios deals with cost - Cost Performance Index. The Cost Performance Index (CPI work package) is based on a ratio of the budgeted cost of the work performed ((BCWP/EV) work package) to that of the actual cost of work performed ((ACWP/AV) work package). The Cost Performance Index (CPI work package) of a project is an efficiency ratio of money being spent on a work package.

(BCWP/EV) work package / (ACWP/AV) work package = CPI work package

Another Earn Value performance ratio deals with the schedule - Schedule Performance Index. The schedule performance index (SPI) is based on the budgeted cost of the work that was performed ((BCWP/EV) work package) and that of the budgeted cost of work schedule ((BCWS/PV) work package). The Schedule Performance Index (SPI work package) is the efficiency of work being completed on the work package.

(BCWP/EV) work package / (BCWS/PV) work package = SPI work package

PROJECT PREDICTIONS

From these calculations, predictions about the future of the work package can be calculated as to the work package's budgeted cost and its scheduled completion. Once again these predictions must be focused on the work package and not summarized on the project level.

The estimated actual cost of the project (EAC project) is calculated by adding together all of the individual estimated actual costs of the work packages of the project (EAC work package). This estimated actual cost of the work package (EAC work package) is calculated by taking the work package budget at completion (BAC work package) and dividing it by the cost performance index of the work package (CPI work package). The work package EAC's are then added together to calculate the project's estimated actual cost for the project (EAC project). An important note for the project management team is that the work packages that have not started maintain a 1.0 cost performance index for the work package unless the project management team has a valid reason for revising the cost performance index of the future work package.

BAC work package/CPI work package = EAC work package
And for future work packages

BAC work package/1.0 CPI (unless possible management adjustment to the CPI) = EAC work package

EAC project = total of ALL project EAC work packages

The estimated scheduled completion of the project (ESC project) is calculated by replacing the original estimated duration of the work package (OED work package) with the revised estimated duration of the work package (RED work package) and recalculating the critical path of the project. The revised estimated duration (RED work package) is calculated by taking the work package schedule performance index (SPI work package) and multiplying it by the original estimated duration of the work package (OED work package).

An important note for project teams is that those work packages that have not started carry a 1.0 schedule performance index for the work package unless the project management team has a valid reason for revising the schedule performance index of a work package.

OED work package/SPI work package = RED work package

John Revere

Earn Value Example

Project Information: *A project has been scheduled for a five week duration with a total budgeted cost of $500,000.00. Work packages have been established and the total cost and duration for each of the work packages have been identified.*

Work Package		Dur. Days	Budgeted Cost	Weekly Expenditure Forecast × $1,000				
				1	2	3	4	5
A	Excavation	5	25k	25				
B	Structural Foundations	5	50k		50			
C	Walls & Roof Structures	12	125k		50	50	25	
D	Interior Work	10	225k			125	100	
E	Site Landscaping	10	50k				25	25
F	Move-in Exp.	5	25k					25
	Budget		500 k	25	100	175	150	50

The critical path for the project is: *A-B-C-E = 32 days*

*****

Project Status : Data date is the end of week 3 of the project.

It is the end of week three and the project work packages have been updated. Some work packages have been completed and others still in progress of being completed.

Work Packages	BAC	%	BCWP	BCWS	ACWP	SV	CV
Excavation	25	100	25	25	30	0	- 5
Structural Foundations	50	100	50	50	60	0	-10
Walls & Roof Structures	125	20	25	100	55	-75	-30
Interior Work	225	50	113	125	130	-12	-17
Site Landscaping	50	0	0	0	0	0	0
Move-in Exp.	25	0	0	0	0	0	0
Budget	500	270	213	300	275	-87	-62

1. Excavation work package has been completed as planned but the actual cost was $5,000 higher.

2. Structural Foundations completed but with an overrun of $10,000.

3. Walls & Roof Structures are 20% complete at this time but should have been 80% (100/125) and $75,000 of work behind schedule, and the cost of the work that has been performed to date is $30,000 over budget.

4. Interior Work has started and is 50% complete but should have been 56% (125/225) at this time and is $12,000 worth of work behind the planned schedule. The cost of the work performed is $17,000 over budget.

5. The remaining work at this time has not, and was not planned to be started.

John Revere

Project Performance Analysis

In general we have a project that is $87,000 worth of work behind schedule and $62,000 over budgeted at this time. This was calculated by the following formulates;

$$(BCWS \ / \ PV) \ project = 300$$
$$(BCWP \ / \ EV) \ project = 213$$
$$(ACWP \ / \ AV) \ project = 275$$

SV project = BWCP project - (BCWS / PV) project = 213 - 300 = -87k

CV project = (BCWP / EV) project - (ACWP / AV) project = 213 - 275 = -62k

WORK PACKAGE ANALYSIS – COST

Cost Performance Index CPI (efficiency of money spent)

Work Packages	BAC	(BCWP)	(ACWP)	CPI	EAC	Complete
Excavation	25	25	30	0.833	30	100%
Structural Foundations	50	50	60	0.833	60	100%
Walls & Roof Structures	125	25	55	0.454	275	20%
Interior Work	225	113	130	0.869	259	50%
Site Landscaping	50	0	0	1.0	50	0%
Move-in Exp.	25	0	0	1.0	25	0%

EAC Project = 699

CPI Project = 0.715

WORK PACKAGE ANALYSIS - SCHEDULE

Schedule Performance Index - SPI (for our example we will assume a direct relationship between duration and performance of money spent and focus is on the work package level).

Work Packages	OED	(BCWS)	(BCWP)	SPI	RED	Complete
Excavation	5	25	25	1.00	5	100%
Structural Foundations	5	50	50	1.00	5	100%
Walls & Roof Structures	12	100	25	0.25	48	20%
Interior Work	10	125	113	0.90	11	50%
Site Landscaping	10	0	0	1.00	10	0%
Move-in Exp.	5	0	0	1.00	5	0%

Schedule Analysis : The original critical path for the project was:

A-B-C-E = 32 days.

The critical path has increased 36 days for a total 68 days. This is based on the poor performance of work package C.

OPERATIONAL SCHEDULE ANALYSIS

Operational schedules primarily focus on the management of project work package resources. Typically those managing the actual resources of the work package may not even know the cost associated with a project work package. Those managing operational schedules should be aware of the resource hours associated with the work package activities and their tasks. The focus of earn value analysis for operational schedules is on resource hours other than cost as in the project earn value analysis.

The first step in operational earn value analysis is to analyze the project work package(s) assigned to the operational management team. The specific operational management team must then identified the specific start and completion dates for all work packages assigned to the organization.

For every operational work package the operational management team breaks these packages down into work activities and then lower into work tasks that a forecast pattern of resources over the life of the work package can be calculated; this information forms the resources operational baseline. This budgeted work hour schedule (BWHS) forms the foundation of the earn value analysis for the operational schedule.

If the operational management team was to take a snapshot anytime during the execution of the operational schedule, they would be able to identify the total resource hours of the work that should have been completed at that point in time (BWHS) and compare it with the budgeted work hours that have been performed (BWHP). The differences between what was to be completed and what has been completed would be an operational schedule variance (OSV).

Operational earn value analysis is performed at the lowest controllable level of the operational schedule - usually the work activity. It should never be performed at a total operational schedule level; this leads to the summation generalization, which leads to improper summary analysis of the operational schedule performance.

$$BWHP \; activity - BWHS \; activity = OSV \; activity$$

A review of the difference between the budgeted resource hours of the work performed (BWHP) and that of the actual resource hours of the work performed (AWHP) would give the project management team a resource variance of the work hours performed (RHV) at that point in time.

$$BWHP \; activity - AWHP \; activity = RHV \; activity$$

From this information the project team can make an estimation of the future resource hour demand and schedule based on the performance to date. The resource performance index (RPI) is based on a ratio of the actual work hours performed (AWHP) to that of the budgeted resource hours of work performed (BWHP). The resource performance index (RPI) of a project is an efficiency ratio of resource hours being spent on the project.

$$BWHP \; activity \; / \; AWHP \; activity = RPI \; activity$$

The operational schedule performance index (OSPI) for the operational schedule is based on the budgeted resource hours of the work that was originally scheduled (BWHS) divided by the budgeted resource hours of work that has been performed (BWHP). The operational schedule performance index (OSPI) is an efficiency of work being performed on the operational schedule.

$$BWHP \text{ activity} / BWHS \text{ activity} = OSPI \text{ activity}$$

OPERATIONAL PREDICTIONS

From these calculations predictions about the future of the operational schedule can be made as to the budgeted resource hours and it's scheduled completion. Once again, this prediction must be made at the lowest controllable operational level, usually the work activity.

The total revised estimated resource hours for the work package (RERH work package) is calculated by adding together the individually revised estimated resource hours of the work package activities (RERH activity).

$$RERH \text{ work package} = RERH \text{ activity} + RERH \text{ activity} + \text{etc.}$$

The individual work package acidity revised estimated resource hours (RERH activity) are calculated by taking the work package activity budgeted resource hours (BWHS activity) and dividing it by the activity resource performance index of the work activity (RPI activity).

$$BWHS \text{ activity}/RPI \text{ activity} = RERH \text{ activity}$$

The various work package activity's RERH's are then added together to formulate the total work package revised estimated resource hours.

An important note for operational management is that those work activities, or any of their tasks that have not started receive a 1.0 resource performance index (RPI) unless management has a valid reason for revising the resource performance index. Management may substitute this analysis with a complete re-estimate of the work by the individual(s) doing the work, if time allows.

The revised estimated resource hours (RERH operational) for the operational schedule is calculated by adding together the individually revised estimated resource hours of the work packages (RERH work package).

RERH operational=RERH work package + RERH work package + etc.

The estimated scheduled completion period for the operational schedule (ESC operational) is calculated by replacing the original estimated duration of the work package (OED work package) with the revised estimated duration of the work package (RED work package) based on work package activity performance and then recalculating the critical path of the operational schedule. The revised estimated revised duration (RED work package) is calculated by taking the individual work package's activity resource performance index (RPI activity) and multiplying it by the original estimated duration of the activity (OED activity).

RPI activity * OED activity = RED work package

John Revere

ISSUES CONCERNING EARN VALUE

Cost Issue

Earn Value gives management a quick snapshot of the project status. Having current and accurate cost information is required for accurate cost variance analysis and estimated cost forecast. This instantaneous financial information rarely is available to the project management team. "Current" means that the data date of information must be included in reporting so that management will not inadvertently rule on a problem or issue.

Accuracy of financial information is always an issue, since resources to investigate project financial information is rarely supported by upper management. The project management team must have resources available to investigate all charges to the project for accuracy of the charges applicable to the period being analyzed.

Another accounting issue is that financial areas of an organization may delay payment to supplemental contractors, suppliers of materials, and suppliers of resources. A delay of payment to these firms may be 30, 60, even 90 days and therefore can cause errors in cost performance analysis and must be excluded from the Earn Value Analysis by the project team.

Another issue for management is when the financial accounting area of an organization prepays invoices to receive a payment discount. Once again this may warp the actual cost to date thus impacting the cost performance analysis of the project.

Schedule Issues

The earn value analysis of schedule does not specifically address critical path performance. Management must closely monitor the critical path and near critical path events to determine the schedule completion date. It is highly suggested that management perform Earn Value analysis on the critical path and near critical path work events to improve their ability to judge project performance. This analysis should include SPI and OSPI calculations on event durations.

Resource Issues

For proper resource earn value analysis, one must have a means of accurately acquiring resource hours in a timely manner. This usually requires a separate program from that of the project, which may not be ideally suited for the project's resource collection requirements.

The management team also may not have had identified support staff to maintain the labor collection program or its cost to the project, and this may be an issue.

Easy of use of labor collection may be an issue if operational resources find it difficult to use the labor collection program.

A labor collection program should have a means of allowing revisions to the resource estimate, and a revised schedule completion date, which are a minimum requirement.

FLOAT ANALYSIS

A simple way to measure effectiveness of schedule performance is to monitor event float. By reporting periodically on every event float associated with every work package, the project team can monitor abnormal changes in work execution. The project management team member responsible for monitoring the work event, which must seek resolution and possibly revise the project or operational schedule, must investigate abnormal differences between periodic updates.

CRITICAL PATH ANALYSIS

The key indicator in determining if a project will be completed on time is to monitor the critical path of the project. To be successful, the project management team must be aggressively proactive in its management of critical path work events. Any critical path event then estimated will cause the project to be completed later than planned. The project team must monitor the critical path events prior to their execution, during the execution, and immediately after for completion fallout, which deals with quality of the work performed and is concerned if the event was completed earlier than scheduled and if the succeeding critical activity may be effective.

It is also required that the project team monitor near critical path events since any change of the project event performance duration, where a work event is completed later than the originally scheduled date, may alter the originally scheduled critical path. The selection of near critical path work events to be monitored by the project team should be identified prior to the project commencement. It should be noted that crashing and fast tracking project events to shorten the project schedule will normally increase the number of near critical and critical work events and therefore more events for the project team to closely monitor.

RESOURCE ANALYSIS

The monitoring of project resources is a prime requirement for project success. Even with documents of understanding or contractual agreements that firms may have with the project team, things happen like: higher priority projects, layoffs, reduction in labor pools due to other external projects, sickness of key employees, employee resignation, unionization of employees, yearly negotiated contracts, and financial reasons, are just some of the reasons that may be a difference between planned and actually applied resources. If resources are not applied according to the anticipated schedule termination of the supplier of the resources may be necessary.

Project Level Performance

At the project schedule level, monitoring of resources is done at an aggregate level. At this level resources are reviewed for number of applied resources associated with work packages. If resources are not being applied as scheduled, work events may, if they are resource based, increase an event duration that may in turn affect the project duration.

Contractual agreements may stipulate that the contractor is to supply adequate resources to complete the project as described by the project team or as defined by the contractor's pre-construction schedule submittal. Internally supplied resources to the project should also be monitored to guarantee that commitments to the project are being fulfilled.

Operational Level Performance

At the operational schedule level, the supplying resource organization needs to closely monitor work events assigned to the organization by the project team. Special effort must be made to monitor the existing and the planned project roster, availability of resources, current utilization rates, and productive and nonproductive rates of the organization. The resource organization must also establish acceptable performance levels for the identified productive resources and inspire these resources to outperform the expected standards of performance of the organization.

Team building exercises, resource performance activities and leadership training can dramatically influence the performance of resources on an operational schedule. It should be noted that at the same time there are just as many deterrents to motivation as well.

The organization supplying resources should report to the project management team, not only work package performance status but also the total resources that are being applied to complete the work packages of the project.

John Revere

SCHEDULE IMPROVEMENT

INTRODUCTION

In order to improve the schedule, the project management team has three major improvement strategies in their arsenal: crashing, fast tracking and containment actions. To choose the best application strategy, the project management team must look at those critical path work events on the critical path. Crashing or fast tracking non-critical path events may add to the cost, but has no effect on the completion date of the project. When analyzing which strategy to use with any critical event, the project team must balance the cost of a strategy with its benefits.

CRASHING

Crashing project events involves adding additional resources to a work event to reduce its duration, and if the work event is on the critical path of the project, reduce the project duration. The addition of resources and its impact to the work event duration is not a straight-line relationship but one of balancing decreased efficiency with these additional resources.

Additional equipment usage must also be considered when adding resources to improve the efficiency of adding resources.

FAST TRACKING

Fast tracking is a methodology that revises the original logical relationships of work events. Reviewing critical work events, that were originally scheduled to be done in series, are now questioned if they can be done in parallel. In most cases some overlapping or paralleling of work can occur. This methodology usually does not increase the cost of the work event or that of the project however; it does increase the risk associated with the logical sequencing of project work events. Any realignment of work event flow must be reviewed by those who developed the original logical work flow of events and assess impact to the entire project for the greatest cost benefit ratio.

CONTAINMENT

Containment is a proactive management tool used by the project team to reduce the impact of future risk events, reduce schedule, reduce cost impacts, or to improve quality of services being offered on the project. By identifying items that could negatively affect the project, proactive tactics can be taken to dramatically reduce the probability of the event or its impact. Containment actions can also be identified and tactics put in place to improve the project performance.

Scheduled containment analysis should be focused on reducing the critical path of the project. By observing elements of the critical path containment, actions like adding additional resources to reduce an event duration can take place. Another containment action could be to add additional equipment or different equipment to reduce scheduled duration of a project work event.

Another common containment action deals with reducing the estimated bids taken on a project, which can be done by spending more money on testing or investigation; this reduces the amount of rework later on in the project's life cycle.

An often missed project containment opportunity involves management resource application. With the addition of a highly skilled management staff, more time can be spent in analysis and study of the project, thus improving the project results. Improved communication and improved documentation are just a few benefits of these additional management resources.

Any containment action used by the project team must have the benefit of the action clearly understood and properly estimated. The containment action must be identified as soon as possible to properly implement its timely execution and its positive impact on the project.

Example: The test boring taken at one thousand foot grid pattern of a site indicates the presences of solid rock between two feet and twelve feet below the ground surface. The new building top of footing elevation is to be at a minimum of four feet below grade or at rock elevation, which is even greater. Bids for rock excavation are expected to be high.

Containment action implemented was to increase the soil boring testing to create a better rock grid pattern based on a five hundred foot grid. Cost of this increased soil boring testing was $25,000.

The resulting bids for rock excavation were $250,000 lower than projected estimates do to this improved grid pattern given to the bidders.

John Revere

GLOSSARY

Activity   An element of a work package which is performed during the course of a project.

(ACWP / AV)   Actual Cost of Work Performed

Application Area   A category of projects that have common elements not present in all projects. Application areas are usually defined in terms of either the project or the type of customer.

AWHP   Actual Work Hours Performed

BAC   Budgeted At Completion - original

Baseline   The original plan, plus or minus approved changes.

(BCWP / EV)   Budgeted Cost of Work Performed/Earn Value

(BCWS / PV)   Budgeted Cost of Work Scheduled / Planned Value

BWHS   Budgeted Work Hours Scheduled

BWHP   Budgeted Work Hours Performed

Change Control   A formally constituted group of
Board   stakeholders responsible for approving or
                            rejecting changes to the project.

Chart of Accounts   Any numbering system used to monitor
                            project costs by category. The project
                            chart of accounts is usually based upon
                            the corporate chart of accounts of the
                            primary performing organization.

Code of Accounts   Any numbering system used to uniquely
                            identify each element of work breakdown
                            structure.

Concurrent   An approach to project staffing that, in
Engineering   its most general form, calls for
                            implementations to be involved in the
                            design phase.

Contingency   The development of a management plan
Planning   that identifies alternative strategies to be
                            used to ensure project success if specified
                            risk events occur.

Contingency Reserve	A separate planned quantity used to allow for future situations which may be planned for only in part, sometimes called known unknown events. Contingency reserves are intended to reduce the impact of missing cost or schedule objectives. Contingency reserves are normally included in the project's cost and schedule baselines.
Contract	A contract is a mutually binding agreement which obligates the seller to provide the specified product and obligates the buyer to pay for it.
Control	The process of comparing actual performance with planned performance, analyzing variances, evaluating possible alternatives, taking appropriate corrective action as needed.
Control Action	Changes made to bring expected future performance of the project into line with the plan.

Cost of Quality The costs incurred to ensure quality. The cost of quality planning, quality control, quality assurance, and rework.

CPI Cost Performance Index

Critical Path In a project network diagram, the series of activities which determines the earliest completion of the project.

CV Cost Variance

Deliverable Any measurable, tangible outcome, result, or item that must be produced to complete a project or part of a project. Often used more narrowly in reference to an external deliverable, which is a deliverable that is subject to approval by the project sponsor or customer.

EAC Estimate at Completion

Earn Value A method for measuring project performance

ESC Estimated Schedule Completion

Exception Report	Document that includes only major variations from plan.
Fast Tracking	Compressing the project schedule by overlapping activities that would normally be done in sequence.
Float (Slack)	The amount of time that an activity may be delayed from its early start without delaying the project finish date.
Functional Manager	A manager responsible for activities in a specialized department or functional area.
Gnatt Chart	A bar chart indicating time and work.
Management Reserve	A separate planned quantity used to allow for future situations which are impossible to predict, sometimes called unknown unknowns.
Milestone	A significant event in the project, usually completion of a major deliverable. It is directly tied to the scope of work of the project.

Network Analysis	The process of identifying early and late start and finish dates for the uncompleted portions of project activities.
OED	Original Estimated Duration
OSPI	Operational Schedule Performance Index
OSV	Operational Schedule Variance.
PCB	Project Control Book - the project documentation record.
Program	A grouping of related projects.
Project	A temporary endeavor undertaken to create a unique product or service.
Project Life Cycle	A collection of generally sequential project phases with names and numbers determined by the control needs of the organization or organizations involved in the project.
Project Management Team	The members of the project team who are directly involved in the project management activities.

Project Plan   A formal, approved document used to guide both project execution and the project control. The primary uses of the project plan are to document planning assumptions and decisions, to facilitate communication among stakeholders, and to document the approved project scope which involves the triple constraint of the project: time, cost and quality baselines.

RED   Revised Estimated Duration

RERH   Revised Estimated Resource Hours

Reserve   A provision in the project plan to mitigate cost and/or schedule risk. Often used with a modifier (management reserve, contingency reserve) to provide further detail on what types of risk are meant to be mitigated.

Retainage   A portion of a contract payment that is held until contract completion in order to ensure full performance of the contract terms.

RHV	Resource Hour Variance.
Risk Event	A discrete occurrence that may affect the project for better or worse.
Risk Identification	Determining which risks events are likely to affect the project.
Risk Management	A subset of project management that includes the processes concerned with the identification, analysis, and response to project risk. It consists of risk identification, risk quantification, risk response development, and risk response control.
Risk Quantification	Evaluating the probability of risk event occurrence and event.
Risk Response Control	Responding to changes in risk over the course of the project.
Risk Response Development	Defining enhancement steps for opportunities and mitigation steps for threats.

RPI   Resource Performance Index

Scope   The sum of the products and services to be provided as a project.

Scope Change   Any change to the project scope. A scope change almost always requires an adjustment to the project cost, schedule, or quality.

SPI   Schedule Performance Index

Stakeholders   Individuals and organizations who are involved in or may be affected by project activities.

SV   Schedule Variance

Workaround   A response to a negative risk event. Distinguished from contingency plan in that a workaround is not planned in advance of the occurrence of the risk event.

Work Breakdown Structure (WBS)   A deliverable -oriented grouping of project elements which organizes and defines the total scope of the project. Each descending level represents an increasingly detailed definition of a project component. Project components may be products or services.

Work Package   A deliverable at the lowest level of the work breakdown structure.

John Revere

Project
Scheduling
Lessons

John Revere

Lesson 1

In this lesson all event relationships are FS (Finish to Start).

Calculate the Early Start, Early Finish, Late Start, Late Finish and Event Float to determine the critical path of this project.

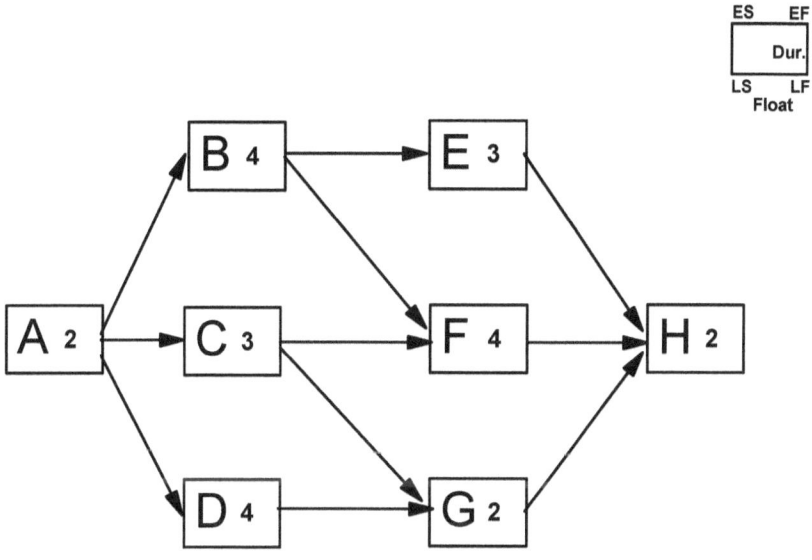

John Revere

Lesson 2

In this lesson all event relationships are FS (Finish to Start) or FS with a lag period.

Calculate the Early Start, Early Finish, Late Start, Late Finish and Event Float to determine the critical path of this project.

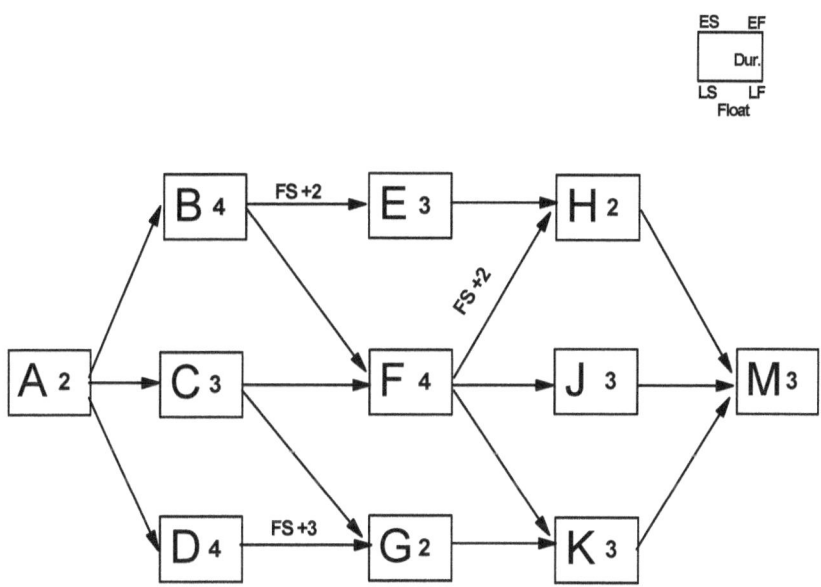

John Revere

Lesson 3

In this lesson all event relationships are FS (Finish to Start) or FS with a lead period.

Calculate the Early Start, Early Finish, Late Start, Late Finish and Event Float to determine the critical path of this project.

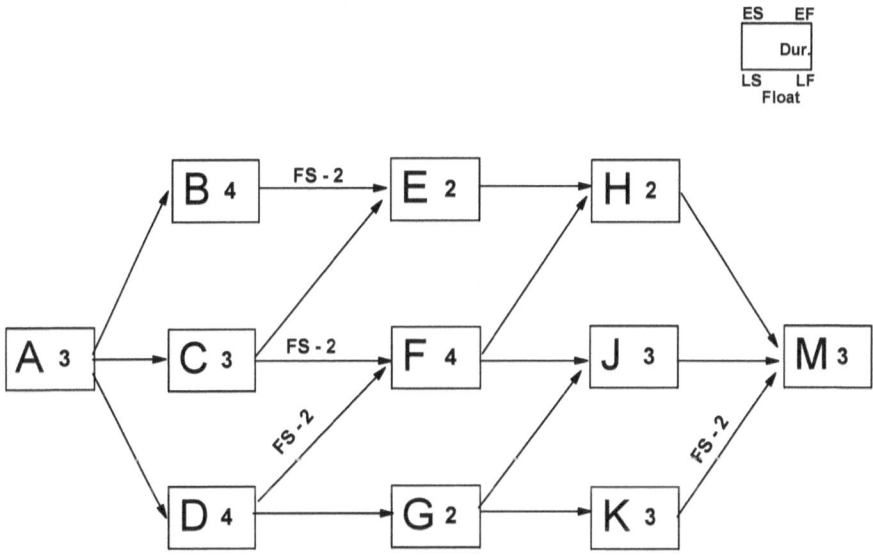

John Revere

Lesson 4

In this lesson all event relationships are FS (Finish to Start) or FS with a lag or a lead period.

Calculate the Early Start, Early Finish, Late Start, Late Finish and Event Float to determine the critical path of this project.

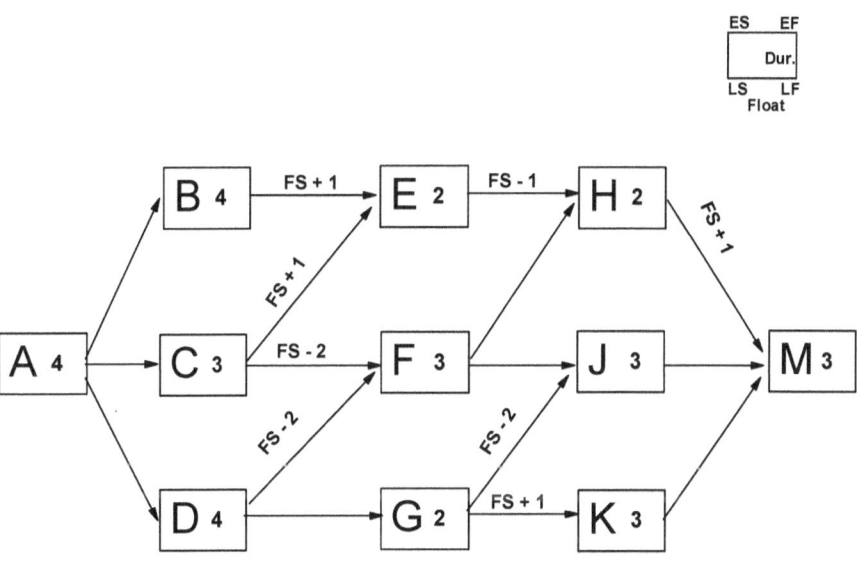

John Revere

Lesson 5

In this lesson we have FS, SS and FF relationships. Some of these relationships have a lag or lead associated with them. We also have an imposed end date of 16. Good Luck!

Calculate the Early Start, Early Finish, Late Start, Late Finish and Event Float to determine the critical path of this project

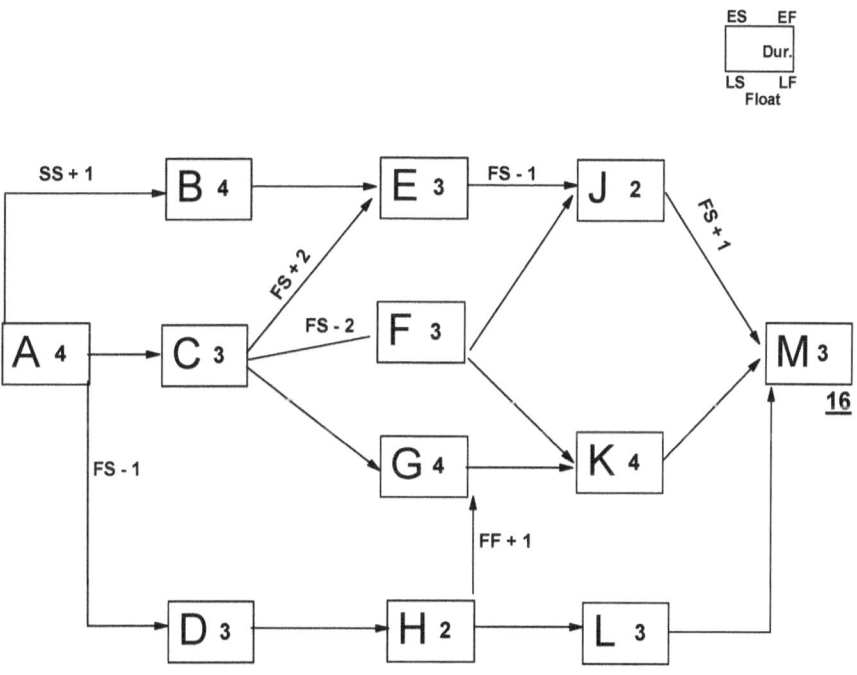

John Revere

Project
Scheduling
Lesson - Answers

John Revere

Lesson 1

Answer

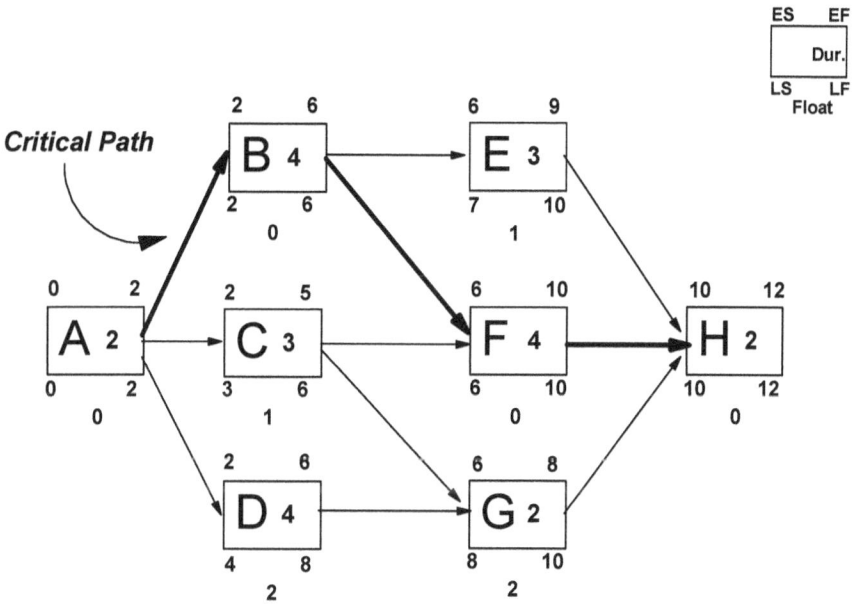

John Revere

Lesson 2

Answer

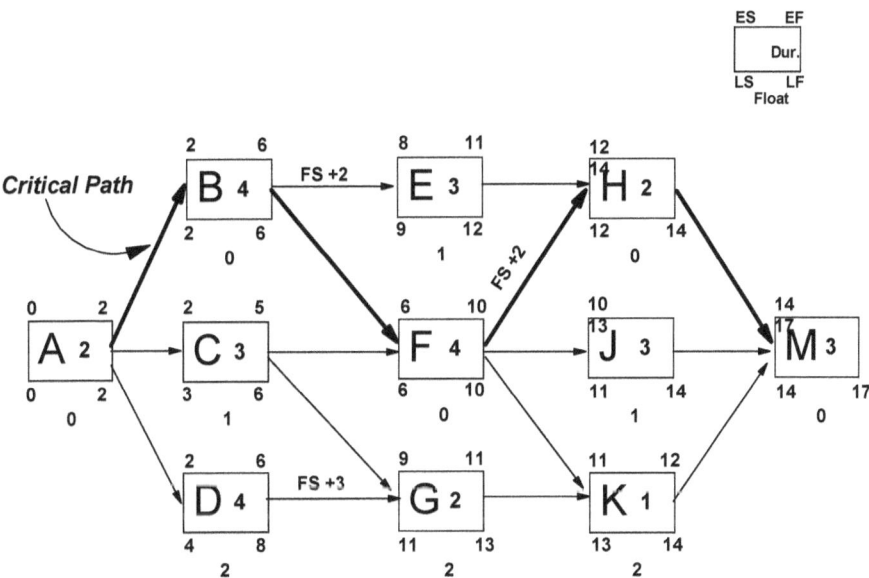

John Revere

Lesson 3

Answer

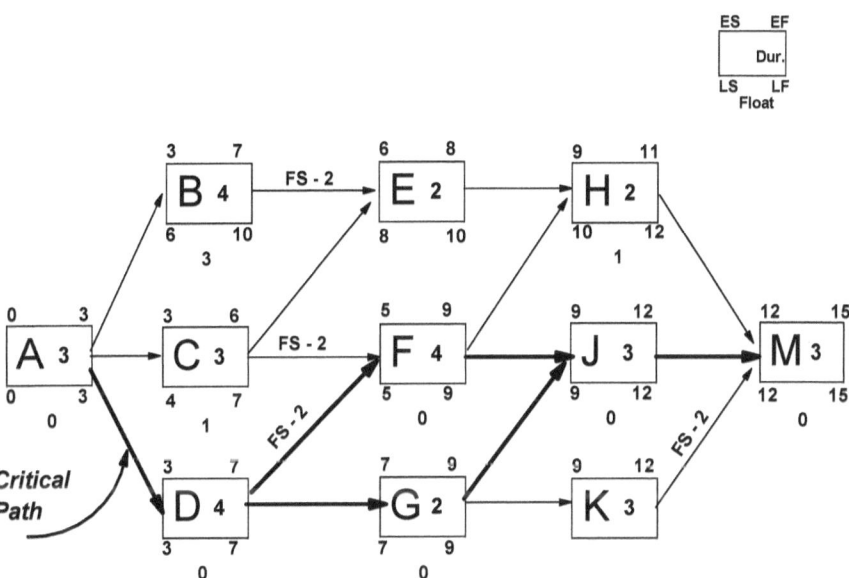

John Revere

Lesson 4

Answer

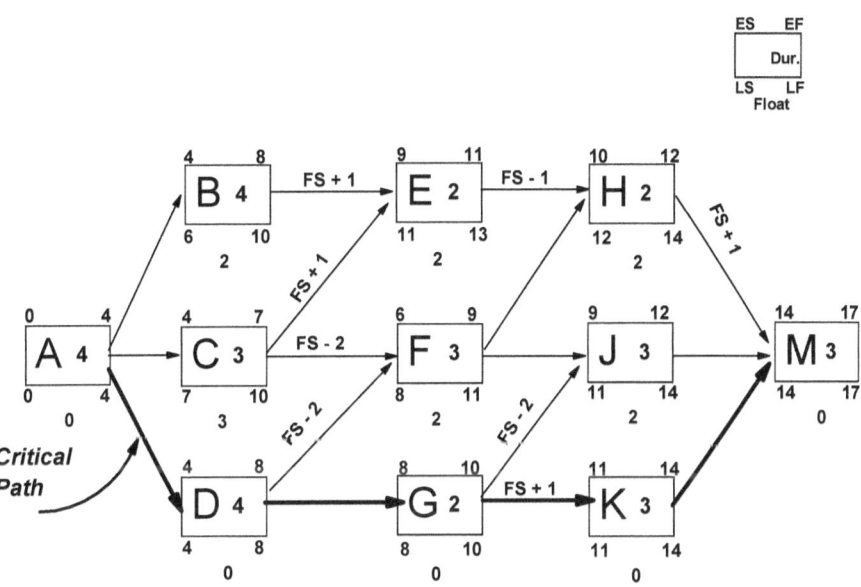

John Revere

Lesson 5

Answer

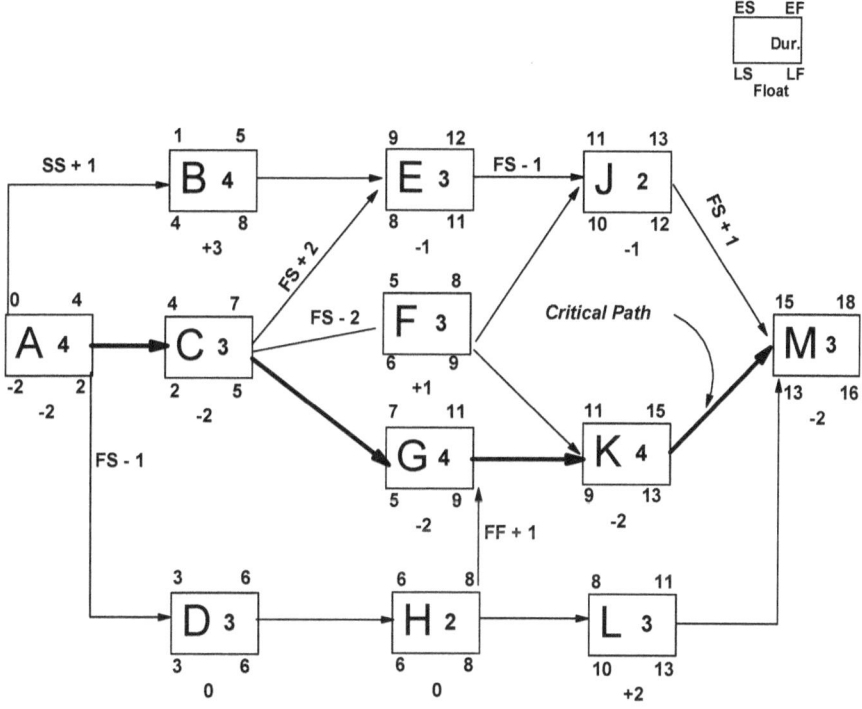

John Revere

More Books By Revere

Construction Risk

Project Administrative Manual

About the Author:

John Revere has worked with Fortune 500 companies (IBM, Best Buy, 3M, Kraft Foods, Motorola, AT&T, Northwest Airlines, Excel Energy and more) to become more profitable on: new product development, construction management, project management, and other customer related service projects. With over three billion in real project experience, he is a leading world educator on the application of performance management.

Revere has a degree in; A.S. in Architecture, B.S. in Construction Management, and an MBA in Management. He has been a member of the Project Management Institute since '87 and is a certified Project Management Professional (PMP).

Other Revere textbooks include: "Construction Risk", "Project Administrative Manual", and "The Management of Aircraft Maintenance".

www.ingramcontent.com/pod-product-compliance
Lightning Source LLC
Chambersburg PA
CBHW022014170526
45157CB00003B/1248